To Andrew Carnegie Esqre.

with Mrs. John Elder's kind regards.

5th January 1893

MEMOIR

OF

JOHN ELDER.

A

MEMOIR

ON CIDER

GLASGOW

A

MEMOIR

OF

JOHN ELDER,

ENGINEER AND SHIPBUILDER,
GLASGOW.

BY

W. J. MACQUORN RANKINE, LL.D.,
*Professor of Civil Engineering in the University
of Glasgow.*

Second Edition.

GLASGOW:
Printed at the University Press by
ROBERT MACLEHOSE, WEST NILE ST.
1883.

PREFATORY NOTE.

THE Author of this Memoir desires to express his grateful sense of obligation to the family, friends, and business connections of the late Mr. ELDER, for the ample information which they have supplied to him, and for the documents to which they have given him access.

W. J. M. R.

GLASGOW UNIVERSITY, 1870.

CONTENTS.

CONTENTS.

APPENDIX.

MEMOIR OF JOHN ELDER.

JOHN ELDER was born at Glasgow on the
8th of March 1824. His elementary education was obtained in the High School of
Glasgow. It does not appear that he applied himself to the study of the ancient
classics; but the result of his training in
English scholarship became manifest in after-
life; for in writing and speaking on those
practical and scientific subjects which he understood so well, he showed himself master
of a clear, concise, and energetic style of
expression.

In arithmetic and mathematics he was a
pupil of Dr. Connell, one of the most able
and successful teachers of the time; and
here he at once gave proofs of extraordinary

A

talent and application, carrying off the principal prizes of the class.

In every branch of drawing—an art intimately connected with mechanical science—he was a most successful student.

The studies before mentioned constituted the principal part of his early school education. A constitution naturally delicate prevented him from deriving the full benefit of his attendance at the High School, and from pursuing his studies to any considerable extent at a university. A short attendance at the class of civil engineering in Glasgow College was all the university education which he received. He was fortunate in being educated under the eye of his father, whose extensive information and high capacity were devoted to the training of his son, and under whose judicious advice he prosecuted his private studies with that ardour which was so marked a characteristic of his later years. The scientific knowledge of which he gave proof in after-life was not only varied and extensive, but also complete and exact, and free from the defects in thoroughness and accuracy which so often beset self-taught scholars.

To those who knew him well, and enjoyed the advantage of personal communication with him, it was manifest that his eminence was due not so much to teaching by others as to the fact that he was that rare character—a man of genius; and therefore in a great measure independent of that external control and guidance which are necessary for the training of ordinary students. In other words, his mind was gifted with the faculty of subjecting itself to the systematic labour and discipline which in ordinary cases have to be enforced by academic authority, and with that strong and clear vision which gives the learner the power of finding his way without a guide through all the mazes of science.

He acquired, as his father had done, considerable knowledge and practical skill in music, especially that of the organ.

He served his apprenticeship of five years as an engineer in the works of Mr. Robert Napier, under the direction of his father, working successively in the pattern-shop, moulding-shop, and drawing-office. He was then employed for about a year as a pattern-maker in the works of Messrs. Hick at Bolton-le-Moors, and afterwards as a draughtsman on the works of the Great Grimsby Docks.

Before 1849 he returned to the works of Mr. Napier to take charge of the drawing-office, a most important appointment.

In the summer of 1852, the firm of Randolph Elliott & Co. of Glasgow, well known and of high standing as millwrights, was joined by Mr. Elder as a partner, and undertook the business of marine engineering, a branch which it had never practised before, and which it was now enabled to undertake through possessing a partner with a thorough knowledge of the principles and practice of that department of applied mechanics. The firm then became "Randolph Elder & Co.," and continued so until 1868, when, on the retirement of the other partners, it became simply "John Elder," and shortly afterwards it was changed to "John Elder & Co." About 1860 the firm added shipbuilding to the other branches of its business.

The career of Mr. Elder, as a marine
engineer and shipbuilder, is so closely connected with the application of the compound
expansive steam-engine to the propulsion
of ships, that it now becomes necessary to
introduce a brief explanation of the principles of that class of steam-engines, and a
summary of their history from the time of
their invention.

In every machine a certain quantity of
energy, or power of doing work, is expended, in order that a certain amount of work
may be done. In every machine, and under
all circumstances, the whole work done is
equal to the energy expended; but a part
only of that work is useful, the remainder
being useless, so that the energy expended
in doing it is wasted. For example, in a
pumping steam-engine the useful work con-
sists in raising, in a given time, a certain
quantity of water to a certain height: the
useless or wasteful work is expended in
overcoming friction. The proportion which
the useful work done bears to the energy
expended is called the *efficiency*. In an
absolutely perfect machine, the efficiency

would be represented by unity—but no such machine exists ; and in every actual machine, the efficiency is expressed by a fraction which falls short of unity by an amount corresponding to the energy that is wasted.

In a steam-engine there are several successive causes of waste of energy.

In the first place, the whole of the energy which the fuel is capable of producing by its combustion is not communicated to the water in the boiler, but only a certain fraction of that energy, ranging in ordinary cases from six-tenths to eight-tenths: this fraction is, the *efficiency of the boiler ;* and the amount by which it falls short of unity corresponds to the heat lost by imperfect combustion, by conduction and radiation, and by the high temperature at which the furnace-gas escapes through the chimney.

Secondly, the whole of the energy which in the form of heat is communicated to the water in the boiler, so as to raise its temperature and convert it into steam, is not obtained in the form of mechanical work done by the steam in driving the piston. In fact, it has for some years been known, through the progress of the science of thermodynamics, that the work done by the steam in driving the piston (often called the *indicated work*, because its amount can be registered by a self-acting instrument called the indicator) corresponds to a quantity of energy which has disappeared from the form of heat, being the difference between the heat brought by the steam from the boiler and the heat carried away by the same steam when it leaves the cylinder. That difference, in every case which can occur in practice, is but a small fraction of the whole heat brought by the steam from the boiler, such as a twentieth, or a tenth; and that fraction is the *efficiency of the steam.*

Thirdly, the whole of the energy exerted by the steam in driving the piston is not communicated to the machine which it is the purpose of the engine to drive; for a fraction of that energy, say from an eighth to a fourth, is wasted in overcoming the friction of the engine—the difference between that fraction and unity being the *efficiency of the mechanism.*

Fourthly, when the machine which it is

the purpose of the engine to drive is an instrument for propelling a ship, a fraction of the energy is wasted in agitating the water in which the propeller works, the remainder only being usefully expended in overcoming the resistance of the vessel, and driving her ahead ; and the ratio which the remainder bears to the whole energy expended by the engine in driving the propeller is a fraction called the *efficiency of the propeller*.

The efficiency of the whole combination, made up of furnace, boiler, engine, and propeller, is found by multiplying together the four fractions already mentioned—viz., the efficiency of the boiler, the efficiency of the steam, the efficiency of the mechanism, and the efficiency of the propeller—and is of course a smaller fraction than any of the factors of which it is the product.

The object of improvements in the economy of the marine steam-engine is to increase as far as practicable, consistently with due regard to economy in first cost, each of the four factors of the efficiency.

Judgment, as well as skill, is specially

required in applying to practice in marine
steam-engineering improvements whose ob-
jects are to increase the mechanical efficiency
of the furnace and boiler, of the steam in
the cylinder, and of the mechanism; for
those improvements for the most part tend
more or less to increase the cost of construction; and thus there arises in each
case the commercial question, Whether the
economy in working to be attained by
means of a given increase of efficiency is
sufficient to warrant the additional expenditure? In deciding that question, regard must
be had to many different circumstances—
such as the length of the voyage, the intended speed, the price of fuel, and the
nature of the traffic. For example, it would
be a waste of money and labour to make
elaborately designed boilers and engines of
very high efficiency for vessels intended to
run short trips between places where coal
is cheap and abundant; while for ships
designed to make long voyages, with few
and distant coaling stations, and expensive
fuel, every improvement that increases efficiency may be a profitable investment. It

is not sufficient, then, for success in the
business of marine engineering, that the
engineer should possess merely a knowledge
of the mechanical principles of his art, and
skill in their practical application—for these
qualifications alone might lead him into need-
less expense in the production of a degree of
mechanical efficiency not required by the
circumstances of particular cases ; he requires
to have also a sound judgment regarding
the commercial result of the adaptation of
engines of a given kind to a given vessel,
intended for a given trade.

Those different qualifications are so sel-
dom found united in one man that the
tendency of popular opinion is to regard
them as incompatible, and to look especially
upon the knowledge, skill, and enterprise
which lead an engineer to adopt new or
unusual improvements in practice as being
fraught with danger to his success in business ; and so no doubt they are, unless
regulated by commercial sagacity.

The success of Mr. Elder and of his firm
proved that his commercial sagacity was not
inferior to his knowledge, skill, and enter-

prise, and that his was one of those rare
minds in which was realised that uncommon
combination of talent.

It now becomes necessary to point out
somewhat more in detail the nature of the
improvements which Mr. Elder, by himself,
or with the co-operation of his firm, carried
out in marine engineering ; and as the
most important of these were connected
with the second and third factors of efficiency already mentioned—that of the steam
in its action on the piston, and that of the
mechanism—the circumstances on which that
second factor depends will, in the first place,
be explained.

The expenditure of energy in the form of
heat required in order to produce a given
weight of steam, when the water is supplied
to the boiler at a given temperature, increases
when the pressure and temperature of the
steam increase, but at a comparatively slow
rate. For example, the expenditure of heat
required to produce a given weight of steam
at the pressure of *ten atmospheres* (about 147
lb. on the square inch of absolute pressure,
or 132.3 lb. on the square inch above the

atmosphere), the feed-water being at the
temperature of about 100° Fahrenheit, is
greater than that required to produce an
equal weight at the atmospheric pressure, in
the proportion only of 1.04 to 1, or 26 to
25 nearly. Hence the problem of obtaining
the greatest possible quantity of indicated
work from a *given expenditure of heat* in
producing steam is so nearly identical with
that of obtaining the greatest possible quantity of work *from a given weight of steam*,
that in practice the difference between those
two problems may be neglected.

All mechanical work is done by the ex-
ertion of a force through a space, and is
calculated and expressed as a quantity by
multiplying the mean amount of the force
into the space through which it acts.   In
the case of steam, the space is the distance
through which the piston is driven in a
given time; the force is the excess of the
forward pressure exerted by the steam behind the piston as it comes from the boiler,
and afterwards expands, above the backward
pressure exerted by the steam in front of
the piston while it is being expelled from

the cylinder into the condenser in condensing engines, or into the atmosphere in non-condensing engines. In a non-condensing engine the back-pressure is a little greater than that of the atmosphere, say from 15 lb. to 18 lb. on the square inch. In a condensing engine the back-pressure is less than that of the atmosphere, to an extent depending on the efficiency with which the condenser acts, or on the goodness of the vacuum, as it is commonly called, and ranges in ordinary cases from 3 lb. to 5 lb. on the square inch.

In an expansive steam-engine, the forward pressure exerted by the quantity of steam that is admitted behind the piston at each stroke has two stages in its action—the admission and the expansion. During the admission the steam is coming from the boiler into the cylinder, and it exerts a pressure less than that in the boiler only by the amount required to overcome the friction of pipes, passages, and valve-ports: say about a twelfth of the absolute pressure in the boiler in ordinary cases. The admission is terminated by the *cut-off*—that is, by the closing of the valve which admits

the steam into the cylinder. Then follows
the expansion of the steam which is confined
in the cylinder; and during this stage of its
action it goes on occupying a continually
increasing space as it drives the piston before
it, and exerting a continually diminishing
pressure. The exact law according to which
the pressure diminishes while the steam expands is complicated, and is different under
different circumstances as to heat. For ordinary practical calculations, however, it is
sufficiently accurate to assume the simple
approximate law, that the pressure varies
inversely as the volume, falling to one-half
of its original intensity when the volume is
doubled, to one-third when the volume is
trebled; and so on.

It is obvious that work continues to be
done by the steam in driving the piston as
long as the pressure behind the piston, or
forward pressure, continues to be greater
than the pressure in front, or back-pressure,
exerted by the steam which has already
done its work, and which the piston is ex-
pelling from the cylinder; and hence it
follows, that in order to realise the greatest

quantity of work which the steam is capable of performing, the expansion ought to be carried on until the forward pressure of the steam behind the piston has fallen so low as to be just sufficient to overcome the back-pressure, and that to end the expansive working of the steam at an earlier period of the stroke is to throw away part of the power of the steam.

This statement, however, must be taken with the qualification, that when the excess of the forward pressure above the back-pressure falls below the pressure which is just sufficient to overcome the friction, the work done is no longer partly useful and partly wasteful, but is wholly wasteful; whence it follows that, although in order to obtain the greatest *indicated* work from a given weight of steam the expansion should be continued until the forward pressure becomes just equal to the back-pressure, the greatest *useful* work is obtained by making the expansion cease when the forward pressure is just equal to the back-pressure added to a pressure equivalent to the friction of the engine.

Another obvious principle is, that both

the indicated and the useful work obtained
from a given weight of steam must be the
greater the greater the proportion in which
the forward pressure exceeds the back-pressure. To take an extreme case : If the
mean forward pressure be simply equal to
the back-pressure, no indicated work whatsoever is obtained from the steam ; and if
the mean forward pressure is simply equal
to the back-pressure added to the friction,
no useful work is obtained. Hence the
higher the forward pressure, and the lower
the back-pressure, the greater is the efficiency
of the steam in an engine ; and as the
pressure increases and diminishes with the
temperature, the same principle may be
otherwise expressed by saying that the
temperature of the steam on its admission
ought to be as high as possible, and that
in a condensing engine the temperature in
the condenser, on which the back-pressure
depends, ought to be as low as possible.
In a non-condensing engine, the back-pressure is, as formerly stated, a little above
that of the atmosphere.

The foregoing principle, as applied to the

temperature in a condensing engine, was first distinctly stated by James Watt; and he invented the separate condenser as a means of carrying it into effect.

The pressure at which the steam is admitted is limited only by the strength and safety of the boiler. In his time, Watt was, in common with most other engineers, very cautious in the use of high pressures ; and he relied more on a low back-pressure than on a high forward pressure for the efficiency of his engines. Improvements in the construction of boilers, and experience of their safety under high pressures when properly designed and managed, have caused subsequent engineers to become gradually bolder in the use of such pressures.

In order to realise the greatest theoretical efficiency in the expansive working of steam, the expansion ought to take place in a non-conducting cylinder, with a non-conducting piston. This condition cannot be absolutely realised in practice ; but means may be taken to diminish the loss of efficiency arising from the conducting power of the cylinder and piston until it becomes unimportant.

B

If that loss arose solely from the waste of
heat by its passage through the metal of the
cylinder to the air outside, it would be sufficient for its practical prevention to clothe the
cylinder with bad conductors, such as wood
and felt. But by far the greater part of that
loss arises in a different and more complex
way, which was not thoroughly .understood
until about 1849 or 1850, when the consequences of the disappearance of heat in performing mechanical work were demonstrated.
Until that time it was erroneously believed,
from reasoning based on the hypothesis of
caloric, that a given weight of steam, after
performing work by expansion, contained exactly as much heat as before, and was therefore superheated ; because the quantity of
heat sufficient to keep it in the vaporous
state at the higher pressure was more than
sufficient to produce the same effect at the
lower pressure ; and so strong was the belief
that the statement of it was distinctly laid
down as a fundamental principle in all,
or almost all, writings on the theory of the
steam-engine.

One of the earliest consequences deduced

from the principles of thermodynamics was, that when steam performs work by expansion a quantity of heat disappears sufficient not only to lower the temperature of the steam to that corresponding to its lowered pressure, but to cause a certain portion of the steam to pass into the liquid state. The steam thus spontaneously liquified collects in the form of water in the cylinder; and if the cylinder and piston were made of a non-conducting material, that water would simply be discharged from time to time into the condenser, without causing any waste of heat. But the cylinder and piston, being made of a conducting material, give out heat to the liquid water which adheres to them, so as to re-evaporate it when the communication with the condenser is opened ; and this heat is carried off to the condenser with the exhaust-steam, leaving the piston and the inside of the cylinder at a low temperature, even though the outside of the cylinder should be clothed with an absolute non-conductor. When steam from the boiler is admitted at the beginning of the next stroke, part of it is immediately liquified through the expendi-

ture of its heat in raising the piston and the
inside of the cylinder again to a high temperature, the result being that at the end of
the second stroke the quantity of liquid
water which is re-evaporated and carries
off heat to the condenser, is greater than it
was at the end of the first stroke. At each
successive stroke that quantity augments
until it reaches a fixed amount, depending
mainly on the difference of the temperatures
of the steam at the beginning and the end of
the expansion; and the effect is the same
as if a certain quantity of steam at each
stroke passed directly from the boiler to
the condenser without performing work. In
some experiments lately made the quantity
of steam which thus ran to waste was found
to be greater than that which performed
work; so that the expenditure of steam was
more than doubled.

The remedy for this cause of loss is to
prevent that spontaneous liquifaction of the
steam during its expansive working, in which
the process just described originates; and
that is done either by enclosing the cylinder
in a *jacket* or casing supplied with hot steam

from the boiler, or by superheating the steam
before its admission into the cylinder ; or
by both those means combined. The steam
is thus kept in a nearly dry state, so as to
be a bad conductor of heat ; and the moisture
which it contains, though sufficient to lubri-
cate the piston, is not allowed to increase
to such an extent as to carry away any
appreciable quantity of heat from the metal
of the cylinder and piston to the condensers.

The steam-jacket outside the cylinder was
invented and used by Watt. Whether he
fully understood the nature of its action can
never be known ; for he did not publish any
reason for using it except that of keeping
the steam as hot as possible. Its real action
was certainly not understood by Watt's im-
mediate successors, nor indeed by any one,
until the principles of thermodynamics were
applied to the question about twenty years
ago ; and many engineers, reasoning correctly
from the erroneous hypothesis of caloric, con-
cluded that the steam-jacket was unnecessary,
and abandoned its use. The fact of liquid
water collecting in the cylinder was known,
but was ascribed to "priming" or the carry-

ing of spray from the boiler. The use of the steam-jacket was retained in a few special kinds of engines, such as the Cornish pumping-engines; and in them the economy properly due to high rates of expansion of the steam was realised; but in almost all other engines, and certainly in marine engines, the jacket was abandoned, with this result—that little or no practical advantage was found to result from expansive working when the steam was expanded to more than about double, or two and a half times its original volume; and this became a received maxim amongst engineers, and especially amongst marine engineers, for its truth in the case of unjacketed cylinders was established by practical experience, as well as by experiments made for the purpose of testing it.

The jacketing of the piston, by filling its internal hollow with hot steam from the boiler, was invented by M. Normand of Havre, and introduced into Britain by Mr. Davison at a comparatively recent date after the action of the steam-jacket had been explained according to the principles of thermodynamics, and its use revived in practice.

As far as the theoretical action of the steam on the piston is concerned, it is immaterial whether the expansion takes place in one cylinder, or in two or more successive cylinders. The advantage of employing the compound engine is connected with those causes which make the actual indicated work of steam fall short of its theoretical amount, and also with the strength of the engine and its framing, the steadiness of its action, and the friction of its mechanism.

The force exerted by the steam on the piston of an engine is transmitted by the piston-rod to the moving pieces of the machinery which it drives—such as the connecting-rod, crank, and crank-shaft; and by the bearings of the moving pieces it is transmitted to the framework. It produces straining actions on all those pieces, moving and fixed; and each of them must be made strong enough to bear safely the straining action produced, not by the mean or average force exerted by the steam, but by the greatest force. The mean force which the

steam has to exert on the piston depends
on the power required to do the work of
the engine, and on the mean speed of the
piston; and the greater the rate of expansion, the greater is the inequality between
the greatest force and the mean force, and
the stronger must the engine be made. For
example, when the steam is expanded to
twice its original volume, its pressure during
its admission is about once and a fifth its
mean pressure; when to five times, its pressure during admission is about double of its
mean pressure; and when to ten times, its
pressure during admission is about three
times its mean pressure; so that in this last
example, if the engine is single cylindered,
all parts of the mechanism and framing that
are strained by the force of the steam must
be made three times as strong as they would
require to be in an engine of the same power
working without expansion. That additional
strength involves not only additional cost of
construction, but additional friction, because
of the greater size of the bearings; and thus
the economy of power due to expansion is
partly neutralised.

It was to obviate this disadvantage in the use of high rates of expansion that the earliest form of compound steam-engine was contrived by Hornblower in 1781. That engine was single-acting, and adapted to the pumping of mines; it had two cylinders, standing side by side, and having their pistons hung from the same end of the walking-beam; the larger cylinder was of the dimensions suited for a single-cylinder engine of the same power and speed; but instead of admitting the steam at its comparatively high initial pressure to act upon the large area of the piston of that cylinder, and thus to exert a great straining force, it was admitted in the first place into the smaller cylinder, so as to exert a straining force equal to the initial pressure multiplied by the area of the smaller piston only; and after having done part of its work by expansion in the smaller cylinder, it was transferred to the larger cylinder in a state of increased volume and diminished pressure to complete its action there. The cylinders were called the high-pressure and low-pressure cylinders respectively, and the same terms are still used in describing compound engines.

The same principle of action was applied
by Woolf to engines with Watt's separate
condenser, and to double-acting steam-engines; and, consequently, compound engines
came to be very generally known as " Woolf's
engines."

In Woolf's form of the compound engine,
as well as in Hornblower's, the two piston-
rods are hung from the same end of a walk-
ing beam, so that the forces exerted through
them act in the same direction at the same
time; and the straining actions produced on
the framing and mechanism are those due
to the sum of those forces. The same is the
case in those forms of direct-acting compound
engines for marine purposes in which the
high and low pressure piston-rods are hung
from one cross-head. Hence, although the
straining actions of the two rods are, in a
well-designed engine of the construction just
mentioned, less than in a single-cylindered
engine of equal power, they are not so small
as they may be made to become by causing
the straining actions due to the two forces
to oppose each other. This improvement,
as far as the straining actions on a walking-

beam and its bearings are concerned, was introduced by M'Naught, who hung the two piston-rods from the opposite arms of the walking-beam, so as to make the difference, instead of the sum of their straining forces, act on the main centre. The sum, however, of those forces still acts on the bearings of the shaft in M'Naught's engine, in the direct-acting engines already referred to, and in the forms of compound engine described in Mr. Craddock's treatise on that subject. That book was published in 1847, and contains the descriptions and drawings of compound engines adapted to marine, locomotive, and other purposes, as patented by him at different times from 1840 to 1846.

Craddock's compound engine, as described by him in the treatise just mentioned, is direct-acting. The high and low pressure cylinders, placed side by side, are not exactly parallel to each other, but make a small angle in order to enable the engine to " pass the centre." The two piston-rods are connected with one crank; upon which, therefore, and upon the shaft and its bearings, they exert a straining action, due to the resultant of their forces, which, though not quite, is very nearly equal to their sum.

Craddock's compound engine, as described in his treatise, is further defective through the absence of steam-jackets, which are now known to be essential to the realising of the economy properly due to high rates of expansion; and unless that economy is fully realised, the additional cost and complexity of a compound engine are thrown away.

It is true that in some steamers fitted with Craddock's engines, or engines resembling them, at a later date (viz., in 1858 and subsequently) the straining actions of the pistons were opposed to each other, and the cylinders were jacketed; but this was long after the time at which the proper principles of the construction of compound marine engines had been brought into practical use by Messrs. Randolph Elder & Co.

In 1850 a peculiar form of compound steam-engine called the "continuous expansion engine" was patented by Mr. Nicholson. The pistons of the high and low pressure cylinders drive two cranks at right angles to each other; and the straining action is the

resultant of those due to the forces acting
through the two rods. This form has considerable advantages in certain cases; but it
was not brought into practical use till about
six or seven years later.

It results, then, from the history of marine
steam engineering, that previous to the formation of the firm of Randolph Elder &
Co., the compound steam-engine had not
been successfully applied in Great Britain
to the propulsion of vessels; that compound
engines such as Craddock's had been proposed for that purpose, but had not been
designed so as fully to realise the advantages of that form of engine; that the aban-
donment of the steam-jacket in the practice
of almost all marine engineers had made it
useless, if not wasteful, to employ those
high rates of expansion to which the compound engine is suited; and that as this
practical error originated in an erroneous
theory of the mechanical action of heat,
founded on the hypothesis of substantial
caloric, then universally prevalent, it was not
to be expected that it should be reformed
except by an engineer who had studied and

understood the principles of the then almost
new science of thermodynamics.

Such an engineer was Mr. Elder. He
knew, in common with other practical men,
that when high rates of expansion were
used with a view to economy of fuel, their
economical action was defeated by the gather-
ing in the cylinders of large quantities of
liquid water, which evaporated when the
exhaust-port opened, and carried away heat
to the condenser ; but he had learned also—
what was known to very few practical men
fifteen years ago—that the formation of that
liquid water originated in the disappearance
of heat during the performance of work by
the expansion of the steam, and that the
remedy was to supply the cylinder with
additional heat to replace that which so disappears, by returning to the practice of Watt
and the Cornish engineers, and resuming
the use of the steam-jacket.

Mr. Elder had also mastered a subject
which, before his time, had been almost
wholly neglected, and which even now does
not always meet with the attention that
it deserves—viz., the diminution of the

friction of the engine by causing the forces
which drive the shaft round to balance and
neutralise, as far as possible, each other's
actions on the bearings where the friction
takes place. As an elementary illustration
of this subject, suppose that a shaft is made
to rotate by means of a single force applied
to a single crank-pin. The whole of that
force will be transmitted to the bearings,
and will there produce a pressure which will
cause a certain amount of friction in addition
to that produced by the weight of the shaft.
But if we now divide the force required to
drive the shaft into two equal forces of half
the amount, and apply them in opposite
directions to a pair of cranks exactly opposite
to each other, those two driving forces will
balance each other as regards pressure on
the bearings, and the friction will be that
due to the weight of the shaft alone. It is
impossible in practice to realise this balance
of driving forces with absolute precision, but
an approach to it can always be made. One
of the most important advantages of compound cylinder engines with opposite cranks
is their enabling that balance of driving

forces to be approximately realised ; and that
advantage had been neglected, or very im-
perfectly developed, before Messrs. Randolph
Elder & Co. constructed their marine engines, which in this respect were a great
improvement upon all compound engines previously invented.

The careful attention which Mr. Elder had
bestowed on the friction of engines and the
means of diminishing it, is fully shown in
an unpublished lecture which he delivered
before the United Service Institution in
April 1866. He there takes a practical example of a marine engine, and shows by
detailed calculation how from 10 to 15 per
cent of the whole indicated power of an
engine may be wasted in unnecessary friction
through neglect of proper arrangements for
the mutual balancing of the forces exerted
on the shaft. In fact, he took a more correct
view of the real advantages of the compound
engine than had previously been done by
any practical engineer ; regarding it as a
means, not so much of increasing the indicated power produced by a given expenditure of steam, as of diminishing that waste

of power which causes the effective power to fall short of the indicated power.

In the lecture referred to, Mr. Elder points out under what circumstances it becomes advantageous to employ a compound engine rather than a single-cylinder engine—viz., when the rate of expansion exceeds four. He adds that, should rates of expansion greater than nine be used, it will become advisable to expand the steam in three successive cylinders instead of in two.

Most of the improvements introduced by Messrs. Randolph Elder & Co. in marine engineering were secured by a series of patents, of which the following is a summary —the patents being distinguished by letters. It is also shown which of those patents were taken in the names of both partners, and which in the name of one only.

A. Charles Randolph and John Elder— dated 24th January 1853. An arrangement of compound engines adapted to the driving of the screw-propeller. The engines are vertical, direct-acting, and geared. The pistons of the high and low pressure cylinders move in contrary directions and drive diametrically

opposite cranks, with a view to the diminution of strain and friction.

B. John Elder—dated 28th February 1854.
For an improved arrangement of the parts
of horizontal direct-acting condensing engines for screw-steamers.

C. Charles Randolph and John Elder—
dated 15th March 1856. This describes an
arrangement of compound engines which
was applied with most successful results to a
long series of steamers. There are two diametrically opposite cranks and four cylinders
making a pair of compound engines; the
high and low pressure cylinder of each engine lie side by side in an inclined position,
and their pistons move in contrary directions. This arrangement not only promotes
the balance of driving forces, but enables
the steam to pass from the high pressure
to the low pressure cylinder in the most
direct manner possible, without having to
traverse long crooked passages, as it did in
Hornblower's and Woolf's engines.

The directions in which the cylinders of
the two engines lean are contrary—that
is to say, for example, in a paddle-wheel

steamer the forward engines incline backwards, and the after engines forward; and in a screw-steamer the starboard and port engines lean respectively to starboard and to port, so that their piston-rods make with each other an angle which, in different engines, ranges from 60° to 90°. The whole arrangement is one of the most simple and compact that is possible in a pair of compound engines, and it produces as near an approach to a balance of driving forces as is practicable when each engine has two cylinders only.

An ingenious contrivance for reversing the engine is described, consisting in an arrangement of epicyclic gearing, whereby a loose eccentric is made when required to overrun the shaft until it reaches the position for backward gear.

The specification states fully the importance of providing each cylinder with a steam-casing or jacket to prevent liquifaction : but this is not claimed; for it was not a new invention, but, as has been already explained, the revival of a practice which had fallen into neglect, though essential to the economical use of high rates of expansion.

D. John Elder—dated 29th January 1858. The specification of this patent describes an arrangement of cylinders in the compound engine by which a nearly perfect balance of driving forces is obtained, and not merely a good approximation to such balance, as in the arrangements previously described; and, consequently, it may be regarded as embodying the principles of the construction of steam-engines of which Mr. Elder approved, in their most complete form, calculated to realise the greatest possible efficiency of the mechanism as well as of the steam. There are three cranks on the shaft—two pointing diametrically opposite to the third, which lies between them. Each engine has three cylinders lying parallel to each other and side by side; in the middle is the high-pressure cylinder, whose piston drives the middle crank; at its two sides are a pair of low-pressure cylinders, whose pistons move simultaneously in the contrary direction to that of the middle cylinder, and drive the other two cranks. Thus the resultant of the forces exerted through the two low-pressure piston-rods is not merely contrary in direction, but directly opposed to the force exerted through the high-pressure piston-rod ; and if the rates of expansion in the high and low pressure cylinders are properly adjusted to their dimensions, there is an exact balance of the actions of those forces on the bearings.

When there is only one low-pressure cylinder, as in the engines described under patent *C*, the forces exerted through the two piston-rods may be equal and contrary, but they are not directly opposed, because they are exerted at different points in the shaft ; and hence the balance of driving forces, cannot be quite exact.

Two or more three-cylindered compound engines can be placed at suitable angles of inclination to each other, so as to drive one shaft, as in the arrangement of two-cylindered engines described in specification *C*.

As the three-cylindered compound engine is somewhat more expensive than a two-cylindered compound engine of the same power, it has been used only in certain cases where special economy of power was desired. Its success in practice will be described further on.

In specification D, as well as in specification C, the importance of the steam-jacket is mentioned; but, for the reason already stated, that part of the engine is not claimed.

E. John Elder—dated 7th June 1858. This patent is for a very simple but very important improvement—the making of paddle-floats of plates of iron or steel, bevelled to a sharp edge, instead of thick wooden planks. The broad edges of wooden paddle-floats oppose a resistance to the plunging them into and drawing them out of the water, and the inventor considered that the substitution for them of comparatively thin sharp-edged metal plates caused a gain of from 4 to 6 per cent in the speed of a given vessel with engines of a given power. This invention was perfectly successful in practice, and was applied to several steamers in the course of the year in which the patent was obtained, and it still continues to be put in practice by the firm with beneficial results.

F. Charles Randolph and John Elder— dated 28th April 1859. This patent is for a variety of improvements in engines and boilers, which it is unnecessary to describe in detail. Amongst other inventions, it describes the making of a boiler with two or more uptakes, in order to increase the surface for superheating the steam.

G. John Elder—dated 15th October 1859. This patent relates to details of mechanism for moving slide-valves.

H. John Elder—dated 25th April 1862. This relates to a variety of improvements on slide-valves, so contrived as to give a smaller opening for the admission of steam and a larger for the exhaust; reversing-gear, in which the position of the eccentric is changed when required by the action of a spiral feather on a shaft which is capable of being shifted longitudinally; arrangements for working steam expansively in four successive cylinders; and an improved kind of water-tube boiler.

I. Charles Randolph and John Elder—dated 20th April 1863. Improvements in surface-condensers, provisionally protected only.

J. John Elder—dated 18th November 1863. This patent is for constructing plate-iron floating-docks, so as to be capable of being navigated from place to place by sails and steam. Three such floating-docks were built by the firm, but were not navigated : one was for Java ; another for the French Government, fitted up at Saigon, in Cochin-China ; the third was for a company in Peru. The two latter have been of great service, and are at present in successful operation.

K. John Elder—dated 19th November 1863. This patent is for various modifications in compound engines, and amongst others for a convenient arrangement of the surface-condenser, in which it is divided into two parts, with tubes parallel to the screw-propeller shaft.

L. John Elder—dated 9th July 1866. This also is for modifications of compound engines.

M. Charles Randolph—dated 15th December 1866. This relates to hydraulic or water-tube propellers.

N. John Elder—dated 28th September 1867. For improvements in floating-batteries —a most remarkable and important invention, which will be described further on.

The first vessel fitted with compound

engines by Messrs Randolph Elder & Co.,
was the screw-steamer Brandon. Her engines
were of the kind described in specification *A*.
She made her trial-trip in July 1854, when
her rate of consumption of coal was found
to be about 3½ lb. per indicated horse-power
per hour. It is well known that the lowest
rate of consumption of coal in steamers
previous to that time was about 4 lb. or 4½ lb.
per indicated horse-power per hour; and such,
indeed, is the greatest economy that can be
expected from such rates of expansion of the
steam as are suitable for unjacketed cylinders.

The Brandon was chartered during the
Crimean war as a despatch-boat, and maintained during many years of service the same
economy which she had realised on her
trial-trip.

The second and third ships were the
paddle-steamers Inca and Valparaiso, for
the Pacific Steam Navigation Company.
The engines of the Inca were started in
May 1856, those of the Valparaiso in July
1856. Each of these ships had a pair of
engines of that compound class described
in patent *C*, already mentioned; the cylinders

were jacketed at top and bottom only, and not round the sides.

The first ship in which engines of the same kind had their cylinders completely jacketed was the Admiral, built by Mr. J. R. Napier, and engined by Messrs Randolph Elder & Co. Her trial-trip was made in June 1858; and in October 1858 she was followed by the Callao, built by Messrs John Reid & Co. of Port-Glasgow. The rate of consumption of coal was found to be: In the Inca 2½ lb.; in the Valparaiso and the Admiral 3 lb.; and in the Callao 2.7 lb. per indicated horse-power per hour—a degree of economy never before realised in marine engines; and this was not only obtained on the trial-trips, but maintained during many years' subsequent service at sea. It amounted to a saving of from 30 to 40 per cent of the coal previously burned by steamers of the same class; and it is not too much to say that it was this saving which rendered it practicable to carry on steam navigation on the Pacific Ocean with profit.

The success of the engines of those ships may be held to have conclusively established

the practical value of the principles on which
they were designed; and it was followed by
the construction, by Messrs Randolph Elder
& Co., of a long series of steamers, in which
the same principles, being more fully carried
out—that is to say, with higher initial pressures, greater rates of expansion, and greater
proportions of superheating surface—realised
even greater economy, the regular rates of
consumption of fuel ranging from 2½ lb. to 2¼
lb. per indicated horse-power per hour.

Another natural consequence was the adop-
tion in the practice of other marine engineers
of the same fundamental principles—that is
to say, the use of high rates of expansion
in the engines of vessels intended for long
voyages, together with the means of causing
such rates to realise their proper economy—
viz., jacketing and superheating. In carrying
out these principles, different forms of engine
have been designed by different engineers—
some have devised peculiar forms of the
compound engine, others have preferred that
the whole work of the steam should be done
in one cylinder. In some cases, forms of
engine that had long before been proposed,

but not executed, have been revived and applied to practice. The detailed history of all these inventions and improvements would be very interesting, but it would be foreign to the purpose of the present Memoir.

In 1865 a comparative trial was made by the Government of the performance of three kinds of marine engines, fitted in three of her Majesty's ships—the Arethusa, the Octavia, and the Constance. These three vessels were of nearly similar model, and of nearly equal size—the tonnage of all three being between 3100 and 3200 tons. Each vessel was fitted with engines of 500 nominal horse-power, and with surface-condensers.

There is no reason to believe that the engines of any one of those three ships were in the slightest degree inferior to those of the others in materials or execution, all three being in these respects of the very first order; and the comparison between them must therefore be regarded as showing how the efficiency of the boilers, engines, and mechanism was affected by the principles embodied in their respective designs.

The principal differences were in the construction of the mechanism of the engine. The Arethusa had a pair of single-cylindered direct-acting horizontal trunk-engines, with cranks at right angles, by Messrs John Penn & Sons.

The Octavia had a set of three single cylinders, horizontal and direct, with double piston-rods acting on three cranks, making with each other equal angles of 120 degrees. These were made by Messrs Maudslay.

The Constance had a pair of three-cylindered compound engines, of the construction designated by D in the account already given of Mr. Elder's inventions, and described as giving the closest approximation to a balance of driving forces on the shaft. Thus the engines of the Arethusa had in all two cylinders, those of the Octavia three, and those of the Constance six.

Those three ships started together from Plymouth at six o'clock in the evening of the 30th September, 1865, in order to run by the most direct course practicable to Funchal in Madeira, a distance of very nearly 1100 nautical miles.

For three days the three ships ran a nearly

direct course under steam alone, the Constance and the Arethusa gaining slightly on the Octavia.

The Arethusa then made sail, and ran for three days more under steam and canvas combined, her course diverging to the eastward. During these three days the Constance and the Octavia continued to run a nearly direct course for Funchal, almost wholly under steam, each having made sail for a few hours only. The Constance continued to gain on the Octavia.

On the 6th of October the Constance was 30 nautical miles from Funchal, 130 ahead of the Octavia, and about 200 from the Arethusa —the last-named ship being about 170 miles to the E.S.E. of the direct course from Plymouth to Funchal.

In the course of the same day the engines of the Arethusa and of the Octavia were stopped, as their coal was nearly exhausted, and they ran nearly all the rest of the way to Funchal under canvas alone, making several tacks.

The engines of the Constance were eased on the 6th of October, because of a westerly

gale and head sea, and she arrived at Funchal
on the 7th of October at 3 P.M., the Octavia
on the 9th at 6.45 A.M., and the Arethusa on
the 10th at 5.35 P.M.

Considering that the last two vessels com-
pleted the trip under sail and in stormy
weather, it is obvious that no fair comparison
between their engines and those of the Constance can be deduced from the *total* time
occupied between Plymouth and Funchal.
In the case of the Arethusa, too, the fact of
her having been three days under steam and
canvas combined makes it difficult, if not
impossible, to form a satisfactory judgment
of her comparative economy of power.

A comparison, however, though a rough
one, of the three vessels, as regards the
consumption of coal per indicated horse-
power per hour, may be deduced from the
official return published by the Admiralty of
the power and of the fuel consumed from the
30th September to the 6th October, when the
engines of the Arethusa and the Octavia were
stopped, and those of the Constance eased.
The following is the calculation, with its
results :—

	Arethusa.	Octavia.	Constance.
Hours under steam, 30th Sept. to 6th Oct., inclusive,	134	140	124
Consumption of coal (tons),	228.85	276.74	242.5
Mean consumption per hour (tons),	1.71	1.98	1.96
Mean indicated power,	1052.2	1399.8	1747
Mean rate of consumption (lb). per indicated horse-power per hour,	3.64	3.17	2.51

As regards the *efficiency of the mechanism*, the same return affords the means of comparing in a general way the Octavia and the Constance; the Arethusa being excluded from the comparison because of her having run so long under canvas and steam combined. The principle upon which the comparison is based is, that in similar vessels of equal size, with mechanism of equal efficiency, the indicated power varies as the cube of the speed; and, consequently, that if for two or more similar and equal vessels the cube of the speed of each vessel be divided by the indicated power, the proportions of the quotients to each other will show the comparative efficiency of the mechanism in the different vessels. The following is the calculation for the Octavia and the Constance, with results :—

	Octavia.	Constance.
Time under steam, . . .	140	124
Distance run (nautical miles), .	1051.7	1090.7
Mean speed (knots), . . .	7.52	8.80
Cube of speed (omitting fractions), .	425	682
Indicated power, do., . .	1400	1747
Quotients, ...	0.304	0.39
Proportionate efficiency of mechanism,	100 :	127
Or, .	79 :	100

This may otherwise be expressed by saying, that at the same speed the Octavia would require 27 per cent more indicated power than the Constance, or the Constance 21 per cent less power than the Octavia. This comparison is not to be considered as very precise; because, strictly speaking, it is the mean value of the cube of the speed, and not the cube of the mean speed, that should be divided by the indicated horse-power.

The superior economy of fuel, as compared with indicated power, in the Constance is, of course, to be accounted for by a higher initial pressure and a greater rate of expansion than those used in the other vessels, combined possibly with better jacketing and greater superheating. But the superiority of the Constance over the Octavia in efficiency of mechanism—in other words, in economy of indicated power as compared with effective power—can be accounted for only by the comparative smallness of the friction in the engines of the Constance; and when it is considered that the engines of the Octavia were of a good design and of the best possible workmanship, the comparative smallness of the friction in the Constance must be ascribed mainly, if not wholly, to the balance of driving forces—the result of the arrangement of cylinders and cranks in Mr. Elder's three-cylindered compound engines.

In a previous series of comparative trials of the Octavia and the Constance, each of the vessels made a run of 100 miles at each of the three speeds of 6, 8, and 10 knots, with the following results :—

	Octavia.	Constance.
Rate of consumption of coal per indicated horse power per hour—		
At six knots, ...	1.90	2.31
At eight knots, ...	2.16	1.95
At ten knots, ...	2.58	2.11
Mean of the three trials, . . .	2.21	2.12

	Octavia.	Constance.
Indicated horse-power—		
At six knots, ...	500	399
At eight knots, ...	1247	1046
At ten knots, ...	1633	1483
Proportionate efficiency of mechanism—		

At six knots, ...	{	100 : : : 80 :	125 100·
At eight knots, . . .	{	100 : : : 84 :	119 100
At ten knots, ...	{	100 : : : 91 :	110 100
Mean, ...	{	100 : : : 86 :	115 100

During this series of comparative trials, the two ships appear to have been nearly equal in economy of fuel for a given indicated power. The superiority of the Constance in the efficiency of the mechanism, though smaller than that deduced from the report of the trip to Funchal, is still sufficient to prove a great diminution of friction through the balance of driving forces in the three-cylindered compound engine, and thus to furnish another practical proof of the soundness of Mr. Elder's views respecting the waste of power in the friction of engines, and the means of diminishing that waste.

Although Mr. Elder invented certain forms

of boiler applicable under special circumstances, he did not confine the practice of his
firm to any peculiar form, but adapted the
boilers to the service for which the vessel was
intended.   His opinion on this point is
summed up in the following quotation from
the lecture already referred to: "A judicious
engineer will therefore design different forms
of boilers for different circumstances, the
object being to construct all his work so as
to give the best return to the capitalist that
employs him."

On the whole, however, he used cylindrical
boilers, fired at both ends, more frequently
than other forms, and latterly he used this
form alone.

The same remark applies to superheaters.
The form of superheater which he generally
employed consisted of an uptake passing
through the steam-chest; and he varied the
extent of superheating surface according to
the degree of economy to be aimed at.

As regards condensation, he approved of
the ordinary jet-condenser for fresh-water
navigation, and for trips of moderate length
in salt water.

For long sea-voyages, his firm and he
latterly adopted the surface-condenser, as
being more economical in working, though
somewhat greater in first cost; nevertheless,
the remarkable economy of fuel in the earlier
compound engines made by the firm was
attained without the aid of surface-condensa-
tion.

The power of calculating beforehand the
probable engine-power required to drive a
given ship at a given speed, or the probable
speed at which a given ship will be driven
by a given amount of engine-power, is obviously of much practical value.

It has long been well known that at
moderate speeds the engine-power required
to drive a given ship varies nearly as the
cube of the speed.

About 1844 Mr. Scott Russell discovered
the law that regulates the limits within which
that principle is approximately true—viz., the
speed must not exceed that with which a
wave naturally travels whose length bears
certain fixed proportions to the lengths of
the entrance and run of the vessel; for as
soon as the speed exceeds that limit, the

power required begins to increase more
rapidly than the cube of the speed. Hence
a *moderate speed* for a given vessel may be
understood to mean a speed not exceeding
the limit determined by applying Mr. Scott
Russell's principle to that vessel. A speed
exceeding that limit may be called an
excessive-speed.

Early in 1859, an investigation of the laws
of the resistance of ships, based on experi-
ment and observation, was made by the
author of this Memoir at the instance of
Mr. J. R. Napier, who required it for practical
purposes in shipbuilding ; and it led to the
result that at *moderate speeds*, in the sense
before mentioned, the resistance is chiefly
of the kind called skin-resistance, depending on the friction between the water and
the immersed surface of the ship, and that
the power required to drive her may be calculated approximately by multiplying the
cube of the speed by a constant factor depending on the roughness or smoothness of
the skin, and by a quantity called the *aug-
mented surface*, which depends on the areas
of the various parts of the skin, and on

their positions relatively to the course of
the particles of water that glide over them
—it being always understood, however, that
the figure of the vessel must be such as to
cause the particles to glide smoothly over
her skin, and not to strike or dash against
it, or become broken into eddies or foam.

The first ship to which those principles
were applied, in order to calculate beforehand the power required at a given speed,
was the paddle-wheel steamer Admiral, built
by Mr. J. R. Napier, and engined by Messrs.
Randolph Elder & Co., in 1858, as already
mentioned ; and the result was perfectly successful. The theory on which these principles were based, and the rules for applying
them, were published in 1860. Mr. Elder,
with that ready appreciation of the practical
value of scientific principles by which he was
distinguished, at once made himself master
of the principles, and continued afterwards
to use them in estimating the probable power
required in proposed vessels.

It has already been shown that Messrs.
Randolph Elder & Co. did not confine their
practice to the construction of that form of

compound engine which approaches the
nearest to theoretical perfection, but adopted modified forms suited to the circumstances of particular cases. In addition to
the instances already given, it may be
mentioned that in many merchant˙ screw-
steamers, where simplicity of construction
was aimed at, they used a form of compound engine resembling that already mentioned as having been first proposed by
Nicholson—a form which of late years has
been adopted by many marine engineers.
There are only two cylinders in all—a high-
pressure cylinder and a low-pressure cylinder; they stand side by side, and their
pistons drive two cranks at right angles to
each other; and there is an intermediate
steam-reservoir, believed to have been first
added to this kind of engine by Mr. E. A.
Cowper, into which the steam passes from
the high-pressure cylinder before its admission into the low-pressure cylinder.   In the
engines of this kind made by Mr. Elder,
the reservoir forms an outer cylinder of the
same diameter with the low-pressure cylinder, and surrounding the high-pressure cyl-

inder; the whole arrangement being very compact and simple, though not having the same advantages in point of balance of driving forces and diminution of friction as are possessed in the highest degree by the engines described under Elder's patent *D*, and in a less degree by those described under Randolph and Elder's patent *C*.

There were cases in which, for the sake of still greater simplicity and compactness, it became advisable to dispense with compound engines and high rates of expansion, as not being required under the circumstances, and of such cases the following is an example:—

Between 1861 and 1864, a demand arose for a class of cargo steamers of very shallow draught, capable of running at a very high speed, not for a great length of time, but on occasions of emergency. Five such vessels were built and engined by Messrs. Randolph Elder & Co. They were a very fine model, driven by paddle-wheels, with feathering plate-iron floats, and each of them had a pair of single-cylindered oscillating engines of ordinary form. Their boiler-power was

very great for their size, so as to provide
the means of producing steam with great
rapidity and of high pressure when required
All those vessels attained a speed of from
16¼ to 16½ knots on their trial-trips; and
that speed was not only realised at sea, but
sometimes even exceeded. On one occasion,
for example, when one of them was very
hard pressed, the bold and skilful officer
who commanded her, succeeded, by an alter-
ation of trim, in increasing her speed to 17
knots, and thus enabled her to escape from
imminent danger.

' The firm of Randolph Elder & Co. was
dissolved by the expiration of the copartnery
on the 30th of June 1868, having existed for sixteen years. During that period
the firm had made 111 sets of marine steam-
engines, whose aggregate nominal horse-
power amounted to 20,145; they had built
106 vessels, whose aggregate tonnage amounted to 81,326; and they had also constructed
three floating-docks. After the dissolution
of the partnership, the business was carried
on by Mr. Elder alone.

The following statement of the quantity of

work executed during the time which elapsed
from the dissolution of the partnership till
the end of the year 1869, shows that the
business had become one of the greatest
of its kind in the world : Number of sets
of engines made, 18 ; aggregate nominal
horse-power, 6110 ; number of vessels built,
14 ; aggregate tonnage, builders' measurement, 27,027.

The number of workmen employed in the
engine-work and shipbuilding yard was about
four thousand. Mr. Elder took a strong and
friendly interest in their comfort and well-
being, and was regarded by them with
corresponding respect and gratitude as an
employer who was just and kind, as well
as able. Amongst other acts of his for
their benefit, he, about half a-year before
his lamented death, promoted the estab-
lishment of an accident fund, by undertak-
ing to contribute to it monthly a sum
equal to that which the workmen should
raise by subscription amongst themselves,
the result being that the income of the
fund is about five hundred pounds a-year.
The fund is managed by a committee partly

appointed by the firm from amongst the foremen, and partly elected by the workmen.

Besides the lecture to which reference has already been made, the views of Mr. Elder on marine engineering are set forth in three papers, which were read respectively to the British Association at Leeds in 1858, at Aberdeen in 1859, and at Oxford in 1860, and printed in the Transactions of that body.

Another lecture, delivered by Mr. Elder to the United Service Institution on the 25th of May 1868, and printed in their Journal, relates to a very remarkable invention, that of circular ships of war. His knowledge of the laws of the resistance of the water to the motion of vessels, led him to the inference that a ship with a hull of the form of a very flat segment of a sphere, like a floating saucer or watch-glass, would require little or no additional power to drive her at a moderate speed, beyond that which is required to drive at the same speed a vessel of equal displacement and of the ordinary form. He tested this conclusion by experiments on models of about five feet in diameter, and found it to be correct; and although at first sight it may seem paradoxical, its soundness will be understood when it is considered that the stream-lines, or lines of motion of the particles of water as they glide over the bottom of the vessel, are, in the case of a flat spherical segment, of a fine form, being either exactly or nearly arcs of circles of a radius equal to that of the sphere. Mr. Elder proposed that a vessel of this form, protected by a belt of armour, and by a deck of sufficient strength, should carry a circular turret suitably armed with guns, and should be provided with a system of submerged propellers, either of the screw or of the hydraulic kind, so arranged as to drive her in any direction, and, when required, to make her turn about her centre, thus dispensing with the necessity for any separate means of making the turret rotate. The probabilities in favour of the success of this invention are so strong, that a trial of it on a practical scale is much to be desired.

Mr. Elder was for four years a captain in the First Lanarkshire Artillery Volunteers; but the multiplicity of his business engagements at length made it impracticable for him to continue to hold that command.

In April 1869, at the annual meeting of the Institution of Engineers and Shipbuilders in Scotland for the election of office-bearers, Mr. Elder was unanimously elected President of that body; and its members looked forward with intense interest to the opening address which he would have had to deliver at the commencement of the session 1869-70. But their hopes were never to be fulfilled; for his health, which had never been robust, at last gave way, and he died in London on the 17th of September 1869, at the early age of forty-five.

He had been married on the 31st of March, 1857, to Isabella, daughter of Alexander Ure, Esq., of Glasgow; and for about three quarters of a year after his death, his business remained in the hands of that lady as sole proprietrix, and was carried on with undiminished success. It then passed into the hands of other partners, but it still continues to bear the honoured name of JOHN ELDER.

Thus far this Memoir has related chiefly to the intellectual powers and the professional

career of its subject. It is not to be sup-
posed, however, that his mental cultivation
was limited to professional matters. He
possessed a large and varied stock of information on most subjects of general interest;
and with his clear head and excellent judgment, it is certain that in whatsoever pursuit
he had chosen for his main occupation, he
must have risen to distinction. The moral
qualities of his mind were of a not less high
order than his intellectual powers. While
firm of purpose and energetic for every good
object, he was kind, generous, and liberal,
and one of the most truthful, just, and
honourable men that ever lived. As regards
the higher aspects of his character, the com-
piler of this Memoir is fortunately able to
produce the testimony of one whose qualifications to speak on that subject are better
than his own. The following is from a letter
of the Reverend Norman Macleod, D.D. :—

.   .   .   .   .   .

"He was a member of my congregation,
and I knew him well. I have seen him in
all variety of outward circumstances—in the
heyday of his strength, vigorous in mind

and body; when suffering from a painful and lingering illness; when ministering to his venerated father on his deathbed, and to his admirable mother in her sorrow. I know what he was to his wife—loved more than all; and very many know, and never will forget, what he was as a friend; and the better I and others knew him, the more we admired and loved him.

"Mr. Elder was truly a religious man. He was not a man of the slightest pretence in anything. He was far too sincere and truthful for *that*. Nor was he given to express, in any degree corresponding to their reality and depth, his feelings or affections, but was singularly calm, quiet, and undemonstrative. His religion was not, therefore, of that type which too commonly and very easily passes in society under the name, merely because certain opinions are held, and certain stereotyped phrases and shibboleths are made use of. His religion was a *life*, not confined to the church or to Sunday, but carried out every day, in the family, in the counting-house, in society, and in business, manifested in untarnished honour, in the sweetest tem-

per, in gentle words, and in remarkable and
most unselfish considerateness for the feelings
and the wants of others.   Such a religion as
his was the result of head, heart, and conscience dealing honestly with truth, and of
a very simple and genuine faith in the love
to him and authority over him of Jesus
Christ.   It was the deliberate choice of a
strong will, affected by a pure mind, quick
conscience, and affectionate heart.   His character told upon every department of his work-
shop and building-yard.   Every one, from the
oldest to the youngest, *felt* the presence of
the man, and were influenced by his goodness as much as by his genius.   In visiting
the other day his great building-yard, one
of his oldest and most trustworthy men,
speaking of him, said to me : ' I never saw
any one like him, nor expect to see his like
again !   He was so just, so true, so kind to
every one.   Every man trusted him, and
knew that he would do all that was possible
to benefit them in every respect.   He had
many plans for their good, which, alas ! he
was not spared to carry out.   I never heard a·
rough or unkind word coming from his lips.'

E

"His funeral was one of the most impressive sights I ever witnessed. The busy works south of the Clyde were shut, forge and hammer at rest, and silent as the grave. The forest of masts along the river were draped in flags, lowered half-mast in sign of mourning. A very army of workmen, dressed like gentlemen, followed his body—column after column. Respectful crowds lined the streets, as if gazing on the burial of a prince; and every one of us, as we took the last look of his coffin and left his grave, felt that we had left a friend behind us."

APPENDIX.

No. I.

Extracts from Letters of the Rev. W. G. Fraser.

LOOKING back on my brief interviews with Mr. Elder, I always felt he was not, like the old philosopher, so absorbed in his mathematics as to forget more vital interests.

When speaking with me on religious subjects, in his own quiet, clear, flowing, and forcible way, about translating the *facts* of Christ's life into our own lives, the unmistakeable impression was left on my mind that he was actually making this part of his own religion, in endeavouring to improve the temporal condition of those around him. Whatever he did for the bodily comfort of those under him, flowed, I have no doubt, from this living principle rising from the centre of his own spiritual being—a God-given and Christimplanted principle in the soul, leading to imitation of Christ in doing good to the bodies of men.

One could not help feeling, in intercourse with Mr. Elder on matters religious, that what he said was not merely from the unseen region of thought, not mere profession and assertion, but experimental from heart and life. And judging only from conversations with him, in ignorance of his mode of caring for his numerous workmen, I shall be disappointed if there is not some proof in the record of his life of the justness of my impressions, that he was one who had at heart the temporal good of his workmen, and who wished this, as Christ wished to *fill the nets* of those who had toiled all night without success. In the spirit of the master, I should conclude, that he carried out Paul's precept, "Be ye kind and affectionate one to another."

Mr. Elder, though always calm, seemed always cheerful, never morose in conversation, sure to add some point or line of light on the subject of discussion. He was one of those "flowing light-fountains" of general knowledge, of unostentatious Christian principle, as well as eminent engineering skill—"a living light-fountain," which one felt (when they had found it) was both pleasant and profitable to abide under its radiance.

Had John Elder been spared to us, I am certain, from the spirit that leavened his motives, from the power combined with gentleness which characterised him, that he would have contributed large practical help in solving some of the difficult problems that are so often springing up between employers and employed in this country. I remember, after his furlough in Russia, how he contrasted the price of labour there with our higher prices here, and how, in genuine sympathy with the working man, he regretted those strikes as frequently far more injurious and disastrous to the men than to the masters, and how he wished to devise some plan whereby the men might be saved the hardships of standing out so long, and trade be prevented from leaving our shores, which it would ultimately do if strikes increased.

That Mr. Elder had not only the temporal interests of the men at heart, but also their highest moral and spiritual interests, I feel certain, from the way in which he spoke of their doubts in a conversation on the infidelity of the age, and the mode he counselled us and all teachers to adopt in grappling with doubters; the apt illustration being that of Thomas, the doubting disciple, who did not at first believe that most vital and fundamental truth, the resurrection of the Lord Jesus.

Yet the Saviour did not frown upon him as an infidel, nor sneer sarcastically at him, but came down and met him on his own ground, as if he entered into his doubts, and asked him to examine for himself the unmistakeable proofs of the *facts* of his resurrection; whereas, had Thomas been treated coolly, and called hard names for doubting what all the others believed, humanly speaking, he might have turned away in confirmed unbelief.

The great Teacher, however, dealing sympathisingly and gently with Thomas, led him, from unbelief to faith, to exclaim, "My Lord, and my God." In like manner (continued Mr. Elder) we should endeavour to meet all doubters on their own ground, giving them credit for what they do believe, and striving to furnish evidence for what they have difficulty about. In this way many might be saved from the ranks of unbelief. Tennyson's lines in 'In Memoriam,' XCV., were partly quoted :—

> " Perplexed in faith, but not in deeds,
>   At last he beats his music out ;
>   There lives more faith in honest doubt,
>   Believe me, than in half the creeds.
>   He fought his doubts and gathered strength ;
>   He could not make his judgment blind ;
>   He faced the spectres of the mind,

And laid them: thus he came at length
To find a stronger faith his own;
And power was with him in the night,
Which makes the darkness and the light,
And dwells not in the light alone."

   .   .   .   .   .   .

Although it was my privilege and happiness to have those frequent interviews with Mr. Elder, and although we had often a quiet chat on religious subjects, yet I should conclude that generally he was reserved on these matters. Never were they obtruded on the general company; and all that he said on those topics was said in that quiet unostentatious manner that impressed me with the feeling that there was in him a deep realising of eternity as closely connected with time.

   .   .   .   .   .   .

No. II.

PACIFIC STEAM NAVIGATION COY.'S OFFICE,
LIVERPOOL, 21st Sept., 1869.

At a meeting of the Court of Directors held here this day, Mr. Charles Turner, M.P., the Chairman of the Company, presiding, the recent death of Mr. John Elder was brought under notice, and it was unanimously resolved that, having regard to the late Mr. Elder's long and valued connection with the Company, a vote of condolence with Mrs. Elder be recorded, that Mr. Just communicate the same, and express the deep sympathy of the Directors with her under her severe and trying affliction.

PACIFIC STEAM NAVIGATION COY.,
LIVERPOOL, 21st Sept., 1869.

MY DEAR MRS. ELDER,—It is now my duty to transmit herewith an extract from the minutes of the Board to-day, expressing the sincere sympathy of the Directors under your present trying dispensation ; and I feel it due, alike to the memory of your late respected husband and to the Directors, to add, that in his death they recognise the loss of a valued connection and private friend.— I remain, my dear Mrs Elder, yours very sincerely.

WILLIAM JUST.

MRS. ELDER, Elm Park, Govan, Glasgow.

PACIFIC STEAM NAVIGATION COMPANY,
HARRINGTON STREET, LIVERPOOL, 23rd Nov., 1869.

Mrs. Elder, Elm Park, near Glasgow :

DEAR MADAM,—I am instructed by the Directors to inform you that they have this day unanimously resolved that, in recognition of your late husband's services to this Company, in the economy of fuel through the use of his compound engines, one of the vessels now building by the firm for the West Coast Service should bear his name. The vessel last contracted for shall therefore be called the "John Elder."

I am, dear Madam, yours very truly,

WILLIAM JUST.

PACIFIC STEAM NAVIGATION COMPANY,
HARRINGTON STREET, LIVERPOOL, 21*st Oct.*, 1870.

In reply to your inquiry as to the extent and nature of the Company's business connection with the late lamented John Elder, and with the firm of Messrs. Randolph Elder & Co., of which he was the guiding spirit—so far as regards marine steam-engines, I may explain that it began in the year 1856, on the occasion of supplying to the Valparaiso a set of engines on Mr. Elder's compound principle—the second, as I believe, of the class made by the firm; shipbuilding being subsequently added to the engineering business, which together were ultimately carried on by Mr. Elder alone. The Company have built no fewer than 22 steam-ships in that yard, and have been supplied, including those now building, with 30 pairs of the double-cylindered engines. In fact, on account of the advantages in the saving of fuel, which, according to our experience, reaches 30 to 35 per cent, we would not think of any other type of machinery.

As you are no doubt aware, the operations have, up to a recent period, been confined to the west coast of South America, where, in consequence of the high price, economy of fuel is of the first importance. It was during the Russian war, when tonnage for the conveyance of coal hence to the Pacific became so scarce, and the cost of the article abroad was thereby more than doubled for a time, that we were led to inquire into the question of a saving of coal. Mr. Elder was called in and consulted, and the double-cylinder engine adopted, as before mentioned, and with a success far beyond our most sanguine expectations, or the advantages held out by Mr. Elder himself. Indeed I am in fairness bound to admit, that his double-cylinder engines never exceeded the promised consumption, nor fell short of the guaranteed speed. On the contrary, the promised results were always more than realised; and I may add, that such was the progress in improvement in the double-cylinder engines, that the last-delivered vessels surpassed the Valparaiso in the economy of fuel as far as she surpassed the ordinary type of machinery.

A short time before Mr. Elder's death, the Company undertook to carry out a mail service for the Chili Government between Europe and Valparaiso, and he was called on to design and construct four large steam-ships of upwards of 3000 tons and 500 horse-power. Those vessels have been so remarkable as regards regularity in performance of the voyage, a distance of 19,000 miles on the round—the greatest steam line in the world—and economical in the consumption of coal, that the attention of many large steam-ship owners, who had long remained sceptical, has been more particularly attracted to the merits of the compound engine, so that ere long I believe the old type of machinery will be unheard of. For this rapid stride in economy, steam-ship owners are, no doubt, indebted to Mr. Elder; and many successful lines of steamers have been projected which never would have had an existence but for the compound principle; thus carrying out the great idea of not only bringing greater advantages and new pleasures into existence, but so cheapening those that previously existed as to bring them within the reach of many who otherwise could not have enjoyed them: and thus also will Mr. Elder's name be transmitted to posterity as a worthy disciple of Watt.

Speaking from long experience, I can aver that, whether in friendship or business, no man could have been more reliable, or more worthy of confidence.

WILLIAM JUST.

No. III.

INSTITUTION OF ENGINEERS IN SCOTLAND,
SECRETARY'S OFFICE, 67 RENFIELD STREET,
GLASGOW, 27th Sept., 1869.

Mr. J. P. Smith begs herewith to transmit to Mrs. Elder the enclosed excerpt minute of meeting of council of the Institution of Engineers, which he trusts she will kindly receive.   He would at the same time desire to express his own sympathy.

THE INSTITUTION OF ENGINEERS IN SCOTLAND,
with which is incorporated
THE SCOTTISH SHIPBUILDERS' ASSOCIATION,
GLASGOW, 27th Sept., 1869.

(*Excerpt of Minute of Council held on 23rd Sept., 1869.*)

It was unanimously resolved that the council of the Institution formally record their deep sense of the great loss the Institution had sustained by the death of their President, Mr. John Elder, and also of the misfortune which had befallen the profession, in losing in the prime of life one whose skill, energy, and varied attainments had done so much for its advancement.

It was further resolved that the council transmit to Mrs. Elder the expression of their sympathy in her bereavement, with the assurance that Mr. Elder's memory will remain with them associated with all that is to be esteemed for high professional ability, integrity of purpose, and trustworthy friendship.

Extracted from the minutes.

J. P. SMITH,
Secretary.

No. IV.

BURGH CHAMBERS, GOVAN.
12th October, 1869.

Mrs. Elder, Govan:

MADAM,—I have the melancholy satisfaction of transmitting to you the annexed excerpt from the minutes of the Police Commissioners of the Burgh of Govan.—I have the honour to be, Madam, your most obedient servant,

W. M. WILSON,
Burgh Clerk.

"At Govan, and within the Burgh Chambers, the eleventh October, 1869. At a general meeting of the Police Commissioners of the burgh, Provost Thomas Reid in the chair,—

"*Inter alia*, the Chairman officially reported the death since last meeting of John Elder, Esq., one of the Commissioners, and moved,—'That the Commissioners resolve to record in their minutes that by the death of John Elder, Esq., their board has been deprived of a member from whose presence, had health permitted it, and life been spared, their deliberations would have derived invaluable aid and enhanced authority; and that the general community of the burgh honour the memory of a marine engineer of distinguished genius and enterprise, while they lament the loss of a large and beneficent employer of labour, a public-spirited citizen, and a good man; and resolve further, that a copy of the minute be transmitted to Mrs. Elder, with the respectful condolence of the Commissioners upon her irreparable bereavement.'

"The Commissioners unanimously approved of the Provost's motion, and instructed the clerk accordingly."

Extracted from the minutes by

W. M. WILSON,
Clerk.

No. V.

ASSOCIATION OF ENGINEERS IN GLASGOW,
GLASGOW, 17*th Sept.*, 1869.

Mrs. Elder:

DEAR MADAM,—In acknowledgment of Mr. Elder's letter of the 7th Sept. last, accepting the honorary membership of this Association, I am directed by the council to express to you our deep sense of the loss we have sustained by his lamented death, and to express our very sincere sympathy with you in your heavy bereavement.— I am, Madam, your respectful and obedient servant,

WM. GEORGE BOWSER,
Secy., Session 1868-69.

ELM PARK.

No. VI.

Extract of Letter of W. Edward MacAndrew, Esq., of Messrs. MacAndrew & Co.

BOND COURT CHAMBERS,
WALBROOK, LONDON, *Oct.* 22, 1870.

. . . .

In all, we had ten steamers built, and three more engined—thirteen in all—by Mr. Elder. I believe that he built his first screw-steamer for us, and she is still running with most satisfactory results—indeed, our unparalleled success in the steam business, in face of severe opposition, is solely attributable to our connection with Mr. Elder enabling us to effect such economies over our opponents.

Mr. Elder was our consulting engineer as well as the contractor for the work, and everything that he could personally superintend was uniformly successful in its results.

.   .   .   .

Both Mr. Elder and myself were animated by a desire to introduce improvements and economies into naval architecture and marine engineering; and it was a knowledge of Mr. Elder's views in this matter which led to our seeking him in the first place. We joined in experiments, which naturally cost something at first, but were ultimately very successful and pecuniarily advantageous to both firms. . . . I know that had he lived, he would have introduced, at least, as great reforms into naval as into mercantile building and engineering.

I have only to add that Mr. Elder was always most liberal in all matters of contract, and by his constant urbanity and liberal execution of all contracts, commanded a preference over all other builders. His personal work and superintendence were hardly less valuable to the business than his irresistible courtesy and unmistakeable intelligence in going into any matter.

.   .   .   .

He was always ready to give his time to discussing any suggestion, whether made by himself or others, and was not only thoroughly scientific, but eminently practical. Unlike other inventors, he did not overstate results to be attained, nor did he press his inventions on those who were too prejudiced to adopt them. He built steamers and engines of the old style for those who so wished them, and always laid the case fairly before his customers. . . . Naturally, old plans and old ways are preferred by many, and few *could* move as fast as Mr. Elder in evolving or executing improved systems.

.   .   .   .   .   .

W. EDWARD MACANDREW.

No. VII.

Extract from Letter of H. Oliver Robinson, Esq., contractor for the Dutch East Indian Steam-Packet Service.

EDINBURGH, 26th October, 1870.

.   .   .   .   .

This service (I may explain) embraces six

F

lines of intercolonial steam-navigation, centering at
Batavia, the capital, performing regular voyages to
and from the following ports : Singapore (con-
necting with the European lines), Samarang, Soura-
baya, and Cheribon, in Java; Padang, Bencoolen,
and Palembang, in Sumatra; Macassar and Menado,
in Celebes; Amboyna, Banda, and Ternate, in the
Spice Islands; Sinkawang and Bandjermassing, in
Borneo; and requiring for the performance of this,
service at least ten steam-vessels of different sizes
or classes.

Those steam-vessels were required to be specially.
adapted for a tropical climate, and for seas where
"fouling" takes place with a rapidity far exceeding
those of a temperate climate; whilst the high cost
of coals (about £2 per ton) rendered economy of
consumption of the first importance.

When to these conditions is added the shallow-
ness of the coasts, the prevalence of coral reefs and
sandbanks, and the absence of lighthouses and
beacons in this Eastern Archipelago, it will be
readily understood that steam-vessels, of a highly
special adaptation were indispensable to success,
both in a maritime and financial point of view.

It is unnecessary for me here to refer to the
aquaintance I had previously the pleasure to form
with Mr. Elder, beyond saying that from it I felt

the conviction that he possessed in a high degree the talents and experience necessary to aid me in designing those steam-vessels, and in determining the leading points in the construction of the vessels and engines.

To him, accordingly, upon my return from Java in August, 1864, I, as the contractor with the Dutch Colonial Government for this steam-service, applied for this aid; and, with his well-known generosity and kindness, he threw himself unsparingly into the subject, and by our joint labours the working drawings and specifications of the whole of the necessary steam-vessels, with their machinery and boilers, were finally settled.

The relative importance of those different lines of steam-navigation necessarily determined the sizes and powers of the steam-vessels, and *three* classes were fixed upon as follows, viz :—

First class, 1050 tons builders' measurement, and 200 horses' power.

Second class, 850 tons builders' measurement, and 150 horses' power.

Third class, 500 tons builders' measurement, and 80 horses' power.

From the fact that all the ten steam-vessels were required to be out at Batavia ready to commence the service on the 1st January, 1866, Mr. Elder's

firm could only undertake to build and engine four,
and to engine a fifth steamer, being all of the first
and second class; and accordingly, contracts for
their delivery "ready for sea" at certain fixed dates
were entered into with the firm, and were duly and
faithfully performed; and on the trial, the speed and
consumption of coals completely fulfilled the stipu-
lated conditions.

These steam-vessels have now been running
nearly five years, without any perceptible falling-off,
and without requiring repairs to either vessels,
machinery, or boilers—the only matter of regret
being that the whole of the fleet could not have
been obtained from the same source.

It may be of interest to refer here to a few of the
peculiar points of those steam-vessels, which were
considered necessary to adapt them to the service
in question.

Their draught light, with great beam.   The
passenger accommodation all on deck, spacious and
airy, covered by a spar deck.   The rig, schooner,
with taunt masts, and large canvas for the light
winds of the Eastern Archipelago, where typhoons
never reach.   But the engines were more especially
the point to which Mr. Elder devoted his attention,
and on which he showed the great liberality of his
mind.

Owing to the circumstances that those steamers never would have occasion to return to Europe in the ordinary course, and that the Suez Canal was *then* not a fact, it was obviously desirable to have their engines and boilers of the most simple design, whilst the light construction of the vessels rendered it of importance to keep down the weight of the machinery as much as possible. To meet these desiderata, single-cylinder engines of the most economical possible consumption, instead of his own *double-cylinder* engines, were proposed to him by me. This idea he at once entered into, and applied himself to the carrying of it out with his usual ardour, and with such success, that upon the trial-trips of the steamers the consumption of *Scotch* coal was only 2¾ lb. per horse-power per hour (indicated).

In all these matters of engineering and construction, the only partner of the firm with whom I came in contact was Mr. Elder, whom, in addition to his great talents and liberal views in mechanical matters, I found to be exceedingly straightforward, as well as easy to deal with on all financial points.

H. OLIVER ROBINSON.

No. VIII.

THE UNIVERSITY, GLASGOW,
21st October, 1869.

I may be permitted to offer some tribute to the memory of Mr. Elder, with whom I had the pleasure of being long and intimately acquainted.

As my firm acted as the law advisers of the concern of John Elder & Coy., and also of the previous concern of Randolph Elder & Coy., I was brought into frequent communication with Mr. Elder, and was thus enabled to form an estimate of his ability and worth. With regard to his ability, I think that it was of the highest description; indeed, I never met any one connected with the noble profession to which he belonged who manifested greater grasp or clearer perception : nay, I go farther, and say that his intellectual power fairly entitled him to be regarded as endowed with genius. Of his worth, whether I think of him as a man of business or in his private capacity, I can speak in the warmest terms. In all his business relations he was the soul of honour ; in private he was ever the kind and considerate friend ; and never, even amid the engrossments of his own immense business, was he forgetful of the wants or the distresses of others. I conclude with remarking that as he lived honoured and beloved, so he died most deeply lamented ; and his memory will always be cherished by his friends, and also by the great community to which he belonged, and with which his name was so long and so honourably associated.

JAMES ROBERTON.

No. IX.

Letter on part of Employees.

FAIRFIELD YARD, *Sept.* 20, 1869.

MY DEAR SIR,—The employees of our late deceased employer, Mr. Elder, are desirous of showing their gratitude, and the manner in which they esteem his memory, by being granted the liberty of following his remains to the place of interment, or part of the way, in whatever manner the relations of our late worthy employer shall see fit to appoint. They will feel grateful by this boon being granted them, as it may be the last open mark that they shall have the liberty of ascribing to his memory, and their sympathy towards his bereaved wife and relations.—

Your obedient servant,

ALEXANDER NEIL.

MR. LORIMER.

No. X.

Letter on the part of Foremen and Workmen.

GOVAN, *22nd Sept.*, 1869.

DEAR MADAM,—At a meeting of the Fairfield Accident Fund Committee (representing the entire body of the foremen and workmen in Fairfield Shipbuilding Yard) it was unanimously resolved to address to you a letter of condolence expressing our sentiments of heartfelt sympathies with you in being bereaved of your loving spouse, and ourselves deprived of a deservingly esteemed employer.

We would refrain from intruding upon your acute grief at this time, but our feelings constrain us to give unqualified expression to our sincere grief for the irreparable loss which you have sustained.

By this sad calamity we mourn the loss of the most benevolent of employers and the most generous of masters—the community, the loss of the enterprising and important supporter—the benevolent and the Christian, that material aid which enabled them to make provision for the needy—the erring, restrained and advised towards a new life.

By this sad calamity we mourn the loss as a star of the first magnitude in the engineering and shipbuilding system which has suddenly vanished, but whose lustre shall outlive the present generation.

By this sad calamity Scotland has cause to weep for an ingenious and illustrious son, rearing a memorial in the hearts of the people which shall remain untarnished during succeeding ages.

And now that we have confidence that he has gone to his rest, we earnestly desire that this providential visitation may be sanctified to you and to us ; may God the Father be to you the husband of the widow, your stay and protector in all circumstances—God the Son your friend and adviser—and God the Holy Spirit your comforter in your sad bereavement, is the prayer of your sincere sympathisers and faithful servants,

ALEXR. NEIL, *President,*
WM. MILLAR, *Secretary,*
for the Fairfield Accident Fund Committee.

To MRS. JOHN ELDER.

9 781019 132883